Chavan's Model Of Galactic Energy Dynamics (CMGED)

https://doi.org/10.5281/zenodo.15250369

Redefining Galactic Rotation, Redshift, and Collapse through Universal Energy Dynamics (UED)

By **Sandeep J Chavan**

Independent Researcher, Architect of CMGED, CMPED & UED

CMGED RESEARCH REPORT

First edition. April 20, 2025.

ISBN: 979-8230868903

Written by Sandeep Chavan.

Table of Contents

"Galaxies are not held together by mass.They're held in place by consequence." — S.J. Chavan

About the Author

Sandeep Chavan is an independent researcher, author, educator, and thinker whose work bridges the disciplines of physics, cosmology, metaphysics, and human intelligence. With over 100 books to his name, including both scientific frameworks and philosophical reflections, he has pioneered a body of work that seeks to restore structure, consequence, and coherence to our understanding of the universe.

He is the originator of the **Universal Energy Dynamics (UED)** framework, which proposes that all physical phenomena arise not from empty space or abstract curvature, but from a structured vacuum field that governs motion, emergence, and dissolution through intelligent equilibrium. His models replace speculative constructs like dark matter, spacetime curvature, and expansion with logically grounded concepts such as rotational consequence, ripple feedback, and dynamic field regulation.

A teacher by temperament and a scientist by realization, Sandeep's work is rooted in direct observation, lifelong inquiry, and a deep respect for natural truths that require no embellishment. He often credits the night sky, silence, and intuitive observation as his greatest collaborators.

With *Chavan's Model of Galactic Energy Dynamics (CMGED)*, he extends the UED paradigm to the galactic and cosmic scale—offering a framework where galaxies breathe, rotate, and dissolve not chaotically, but harmoniously within the structured fabric of the vacuum.

He believes the universe is not a puzzle to be solved but a rhythm to be understood—and that true science begins when we dare to see again what we have long ignored.

"The laws of consequence are not human constructs. They existed before thought, before time—written in rotation, revealed through silence."

You can explore his broader work—on gravity, time, intelligence, karma, and consciousness—through the UED Research Series, the War of Thoughts dialogue books, and his latest philosophical science titles.

ORCID Profile: https://orcid.org/0009-0005-4190-7164

"To the silence that spins the universe."

To my dear students,

To my children and my family,

And to the great scientists—past, present, and future—

whose restless minds dared to ask,

"What if the universe works differently?"

This work is a tribute to your curiosity, your questions,

and the eternal dance between thought and truth.

Abstract

Chavan's Model of Galactic Energy Dynamics (CMGED) presents a unified field-based framework for understanding the rotational behavior, stability, and structural dynamics of galaxies without invoking hypothetical entities such as dark matter or the expanding universe paradigm. Building on the foundation of Universal Energy Dynamics (UED) and core mechanics of Chavan's Model of Planetary Energy Dynamics (CMPED), this model introduces galaxies as annular rotating systems embedded within Structured Vacuum Energy (SVE), a real energetic medium that governs motion through field tension and consequential equilibrium rather than gravity alone.

The model asserts that a galaxy's angular momentum is conserved not by inertial isolation but by continuous mass-energy redistribution, enabling a constant angular velocity to be maintained across the galactic disc (Ro–Ri structure). Redshift is reinterpreted as a function of rotational frame misalignment—arising from differences in angular speed between galaxies—rather than as evidence of spatial expansion.

Gravitational waves are redefined as radial ripple corrections, emitted from galactic cores or catastrophic events, functioning as equilibrium-restoring signals within the SVE field.

By explaining why constellations appear fixed, why galaxies don't collapse, and how mergers occur as field-based consequence resets, CMGED offers a holistic, mechanically grounded cosmology.

It integrates observed stability, motion perception, time variation, and ripple phenomena into a single, consequence-driven framework.

This report proposes that the universe is not expanding chaotically but rotating harmonically, with galaxies, stars, and planetary systems breathing together in nested, intelligent balance with the field.

Part I: Foundations of the Model

1. Introduction: The Need for a New Galactic Model

Modern cosmology, despite its mathematical sophistication, continues to rely on speculative constructs to explain fundamental phenomena in the universe. The prevailing models—rooted in general relativity, the Big Bang hypothesis, and dark matter theory—have managed to fit observational data only through adjustment, patchwork, and approximation. However, these constructs raise more questions than they resolve, particularly when scrutinized under the lens of physical causality and coherent dynamics.

The dark matter hypothesis, proposed to account for the unexpectedly flat rotation curves of galaxies, assumes the presence of invisible mass that interacts gravitationally but remains undetectable by any direct means. It explains what is observed, but it does not explain why the system behaves in that way from a first-principles standpoint. Likewise, the concept of spacetime curvature provides a geometric interpretation of gravity, yet lacks a tangible medium or structural basis through which motion and mass-energy relationships can be understood mechanically. It replaces physical tension with abstract deformation.

The Big Bang model, with its singular explosive beginning, fails to account for the structured rotational equilibrium of galaxies, the symmetry of spin vectors across vast distances, or the emergence of complex forms in a supposedly chaotic early universe. Furthermore, the assumption of an expanding universe, based largely on redshift data, interprets observed energy dilution as spatial recession without addressing alternative causes rooted in field interaction and resonance.

These limitations demand a rethinking of galactic behavior—one that is not dependent on theoretical placeholders, but rather on observable consequences and structured energy dynamics. The **Chavan Model of Galactic Energy Dynamics (CMGED)** emerges from this need. Rooted in the principles of **Universal Energy Dynamics (UED)** and **Chavan's Model of Planetary Energy Dynamics (CMPED)**, CMGED proposes a field-based cosmology where galaxies are not isolated masses in expanding emptiness, but structured, co-rotating annular systems embedded within a dynamic medium—Structured Vacuum Energy (SVE).

This new model is not only consistent with existing observations, but it also provides mechanical causality, energy balance logic, and testable predictions. By replacing abstraction with structure and theoretical mass with dynamic equilibrium, CMGED opens the door to a cosmology that is intelligent, rotationally stable, and fundamentally observable.

2. Overview of UED and CMPED Principles

The Chavan Model of Galactic Energy Dynamics (CMGED) is a natural evolution of two foundational frameworks developed within the Unified Energy Dynamics (UED) paradigm:

1. Structured Vacuum Energy (SVE) as the universal medium, and
2. Chavan's Model of Planetary Energy Dynamics (CMPED) as the local model for orbital and rotational behavior of celestial systems.

Together, these two models provide the conceptual and mechanical basis upon which CMGED is constructed.

2.1 Structured Vacuum Energy (SVE): The Foundational Medium

In conventional physics, the vacuum is often treated as empty space—a passive backdrop upon which forces act and particles interact. However, UED redefines the vacuum not as emptiness, but as a structured, energy-rich medium.

This Structured Vacuum Energy (SVE) is not a mathematical abstraction but a real, physically consequential field that:

- Carries latent energy and structured tension,
- Governs the formation, interaction, and balance of all matter and motion,
- Enables the emergence of time and space as perceived

phenomena within localized systems.

SVE behaves like a fluid with embedded intelligence. It stores, distributes, and balances energy across scales, from quantum particles to galaxies.

When a system moves or changes within SVE, it does so not freely, but through a feedback mechanism that seeks dynamic equilibrium.

This medium is not curved by mass—it responds to imbalance with realignment, ripple propagation, and structural reconfiguration.

2.2 Consequence-Driven Motion

One of the core principles of UED is that motion is not spontaneous—it is consequential. That is, nothing moves unless there is: An imbalance in energy distribution, A difference in field tension, A necessity to restore equilibrium.

All rotation, translation, and oscillation are not results of inertial continuation but of structured rebalancing efforts within the vacuum.

In this view, angular momentum is not a mystical conserved quantity—it is a consequence of field structure seeking balance through continuous redistribution of energy.

In contrast to Newtonian models, where motion is sustained unless acted upon, the UED framework asserts that motion is always being acted upon—by the field. Hence, what appears stable or constant is actually a highly regulated dynamic process held together by the law of consequence.

2.3 From Planetary Energy Dynamics (PED) to Galactic Energy Dynamics (GED)

The Chavan's Model of Planetary Energy Dynamics (CMPED) established that planetary systems—such as our solar system—are not simply collections of bodies held together by gravitational pull, but co-rotating, energy-regulated systems embedded within local SVE shells.

In CMPED:

- Planets do not orbit the Sun due to gravity alone, but because their mass-energy is tuned to maintain a structured motion relative to the vacuum field shaped by the Sun.
- The Sun itself is not a stationary mass, but a dynamic energy node regulating local vacuum conditions.
- Spins, tilts, and orbital behaviors emerge from attempts to maintain rotational balance, not from random formation.

CMGED builds directly upon this. While PED explains stability at the stellar system scale, Galactic Energy Dynamics (GED) operates at the next level—where:

- Stars, systems, and spiral arms are all nodes embedded in a larger annular ring (Ro–Ri),
- The entire disc of the galaxy co-rotates as a result of embedded energy dynamics in the SVE field,
- Angular momentum is preserved not passively, but by actively rebalancing moment of inertia across the galactic structure.

In essence, GED is PED applied at a higher order of complexity and scale. T he same principles of structured vacuum, consequential motion, and rebalancing hold true, but now they govern the dynamic behavior of stars within galaxies, and galaxies within cosmic structures.

CMGED thus stands as the natural extension of UED and CMPED—offering a self-consistent, mechanically grounded cosmology that replaces dark matter, curved spacetime, and cosmic expansion with real structure, rotational equilibrium, and field-regulated consequence.

3. The Ring Model of a Galaxy (Ro–Ri Structure)

At the heart of Chavan's Model of Galactic Energy Dynamics (CMGED) lies a redefinition of what a galaxy *is*. Contrary to the traditional depiction of galaxies as massive, loosely bound systems held together by invisible gravitational scaffolding (i.e., dark matter), CMGED posits that galaxies are structured, self-regulating, rotating energy rings embedded within the Structured Vacuum Energy (SVE) medium.

This section outlines the Ro–Ri model—the annular shell within which all galactic dynamics occur—and provides a clear mechanical reinterpretation of how galaxies achieve rotational coherence, long-term stability, and self-preservation through field-based equilibrium.

3.1 The Galaxy as an Annular Rotating Structure Embedded in SVE

A galaxy, in CMGED, is not a solid body or a chaotic cloud of stars. It is a fluid-like, field-bound annular system, defined by a rotational shell with a clearly regulated structure. The entire galaxy spins not due to inertial momentum inherited from a Big Bang, but due to a rotational consequence established within a structured field — the SVE.

This annular ring is more than a collection of moving stars. It is a coherent energy shell, with internal field harmonics that maintain:

- Constant angular speed across the disc,
- Distributed mass-energy realignment to preserve moment of

inertia,
- System-wide communication via ripple feedback (perceived as gravitational waves).

This is not orbiting — it is embedded spinning. The galaxy is not held together — it is balanced together.

3.2 Definition of the Rotational Energy Shell (Ro–Ri)

The Ro–Ri structure defines the galactic shell where all observable dynamics occur. It consists of:

- **Ri**: The **inner boundary** of structured rotational field — the closest stable region surrounding the central black hole. This is not a geometric center but the beginning of the dynamic equilibrium zone, where core friction and energy tension begin to shape orbital behavior.
- **Ro**: The **outer boundary** of the structured disc — the point beyond which stars are no longer co-rotating with the main system and field coherence fades. Beyond Ro, stars become part of the galactic halo or migrate toward dissolution into the intergalactic SVE.

Between Ri and Ro, the galaxy exists as a rotational equilibrium zone, where:

- Angular speed ω is maintained,
- Mass distribution is constantly readjusted to keep moment of inertia I stable,
- Energy flows inward (toward Ri) as stars age, collapse, and are absorbed by the central core,
- Ripple corrections (gravitational waves) flow outward to maintain systemic balance.

This Ro–Ri ring is not merely spatial — it is energetic. It defines the *active layer* of the galaxy, where all motion, light, and structural memory reside.

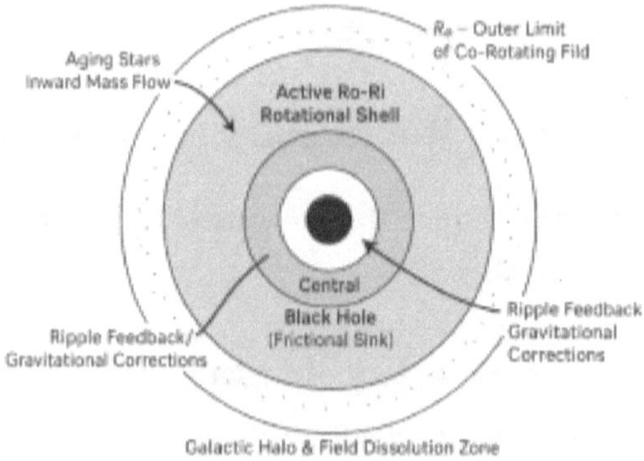

Terminology Snapshot: Core Concepts of CMGED

A compact visual glossary for readers to navigate the dynamic architecture of galactic energy regulation.

Structured Vacuum Energy (SVE)

Definition: The universal energetic medium within which all motion, structure, and balance occur.

Role: Acts as the "field" responsible for rotational coherence, ripple propagation, and energy rebalancing.

Think of it as the invisible scaffolding that holds everything in place—not space, but structured energy.

Ro–Ri Shell (Rotational Energy Zone)

Ro (Outer Radius):

- The **outer boundary** of the active galactic disc.
- Beyond this, stars **lose phase coherence** and enter the galactic halo.

Marks the fading edge of rotational synchronization.

Ri (Inner Radius):

- The **innermost stable boundary** of the rotational shell.
- Field tension begins to spike here near the **central black hole**.

Defines the transition from orbital identity to energetic absorption.

Ro–Ri Shell (In Between):

- The active, dynamic region where all motion, ripple, and regulation occur.
- **Angular velocity** (ω) is regulated.
- **Mass distribution adjusts** to maintain rotational balance.

Ripple Feedback

Definition: Radial energy pulses emitted through the SVE when system imbalance occurs.

Role:

- Maintains equilibrium by correcting angular tension.
- Manifest as **gravitational waves** in mainstream terms.

Field's natural response to imbalance—like a galactic heartbeat.

Consequence-Driven Motion

Definition: Motion arises not from applied force, but from **tension within a structured field.**

Role: Ensures that all energy movement is part of an **equilibrium-seeking process**, not inertia.

Motion is never free—it's always a feedback to imbalance.

Ripple Friction (Core Frictional Sink)

Definition: The condition near the galactic center where incoming mass loses individual orbital energy and is absorbed into the field.
Role:

- Prevents collapse by **converting motion into internal ripple.**
- Analogous to the **black hole** acting as a **rotational balancer.**

Not destruction—structured reintegration.

Rotational Time Base

Definition: The perceived "clock" of a system, determined by its **angular velocity** (ω) within the Ro–Ri structure.

Role:

- Defines how time is experienced **locally** in a galaxy.
- Redshift arises due to **time-base mismatch**, not expansion.

Each galaxy spins in its own rhythm—redshift is the cosmic echo of those differences.

3.3 Core vs. Disc Dynamics

CMGED emphasizes a crucial distinction between the central core and the rotational disc of a galaxy.

The Core:

- Anchored by a supermassive black hole, which in UED is not a destructive singularity, but a frictional sink of energy tension.
- Acts as the consequence regulator — absorbing excess mass-energy as stars and systems migrate inward.
- Releases ripple feedback (gravitational waves) to realign field tension throughout the disc.
- Serves as the final point of equilibrium, where structured energy is returned to SVE.

The Disc (Ro–Ri Zone):

- Contains the bulk of the galactic structure: stars, nebulae, planetary systems, spiral arms.
- Co-rotates with constant angular velocity, regulated not by mass, but by energy field geometry.
- Mass-energy is not static — it migrates slowly inward due to frictional interactions and gradual loss of orbital energy.
- Angular momentum is preserved globally by internal redistribution of matter and spin via dynamic field regulation.

The disc is the body of the galaxy. The core is its breath regulator. And the SVE ring is its skeleton and skin.

This ring model replaces the gravitational mystery of dark matter with an elegant mechanical alternative: A galaxy spins, holds together, and maintains structure because it is a consequence-governed, energy-structured, rotational system embedded in a living vacuum — not because of unseen mass or mystical spacetime warps.

Part II: Core Dynamics of CMGED

4. Angular Momentum and Moment of Inertia Regulation

In classical mechanics, angular momentum is a fundamental conserved quantity for rotating systems, represented by the equation:

$L=I\cdot\omega$

Where:

- L is angular momentum,
- I is moment of inertia (which depends on mass distribution),
- ω is angular velocity.

In the context of galaxies, this equation becomes more than a mathematical identity — it becomes a diagnostic tool for understanding how large-scale rotational equilibrium is achieved and maintained over billions of years.

The Chavan Model of Galactic Energy Dynamics (CMGED) reinterprets this equation within the field-based logic of Structured Vacuum Energy (SVE), demonstrating that the galaxy itself is a self-balancing system where moment of inertia is not fixed but dynamically regulated to preserve angular momentum.

4.1 The Equation $L=I\cdot\omega$ in the Galactic Field Context

In a rotating galaxy, L is not simply the sum of individual star motions. It represents the net rotational consequence of the entire galactic energy ring (Ro–Ri) embedded in SVE.

Classically, if no external torques act on a system, L is conserved. However, galaxies are not rigid objects, nor are they completely isolated. They are field-embedded, multi-scale, energy-regulating structures. Therefore:

- I (moment of inertia) constantly evolves due to redistribution of mass across radial distances,
- Yet ω (angular velocity) appears remarkably constant across the disc,
- Hence, L is preserved not passively, but by active, internal rebalancing mechanisms.

This reveals a field implication: The vacuum field (SVE) exerts a consequential feedback mechanism to ensure rotational equilibrium. It compensates changes in I through corresponding micro-adjustments across the system, maintaining ω nearly constant for the entire galactic shell.

4.2 Why Galaxies Maintain Constant Angular Velocity (ω)

Observationally, galaxies exhibit flat rotation curves — stars at different radial distances from the center rotate with similar tangential velocities, which implies nearly constant ω\omegaω in the inner-to-mid disc region.

According to Newtonian mechanics, outer stars should rotate slower due to lower gravitational pull. But this is not what we see. Instead, the entire disc behaves as if it's rotating as a single energetic unit — not because of gravity, but because of the underlying field regulation.

In CMGED, the constant ω is not coincidental — it is required for systemic coherence in the rotating Ro–Ri ring. The field imposes rotational alignment to:

- Maintain tension symmetry across the annular band,
- Minimize angular shear,
- Stabilize energy flow from outer to inner regions,
- Ensure internal dynamics remain regulated.

Constant angular speed is not a relic of formation — it is a function of ongoing consequence.

4.3 Mass-Energy Rebalancing via Ripple Feedback

The galaxy, in this model, is never truly static. Though it appears stable, its internal field dynamics are constantly engaged in balancing angular momentum through:

A. Inward Migration of Mass-Energy

- Over time, stars lose energy through radiation, frictional interaction with local field structure, and gravitational interaction with other bodies.
- This causes a slow radial drift inward, especially for older stars or collapsed systems, toward the central frictional sink — the black hole.
- This process increases I unless compensated.

B. Ripple Feedback Mechanism (Gravitational Waves)

- As the moment of inertia shifts due to inward migration or star death, the field releases corrective radial ripples — perceived externally as gravitational waves.
- These ripples act like vacuum field updates, rebalancing angular tension and stabilizing ω.
- They also distribute micro-adjustments throughout the ring, realigning mass distributions in a field-wide feedback loop.

C. Spiral Arm Dynamics

- Spiral arms serve not only as structural features but also as energy distributors, where new stars are formed and old ones are funneled inward.
- They help dissipate angular tension, making mass-energy rebalancing smoother and non-destructive.

Ripple feedback is the galaxy's way of communicating with itself. It ensures that mass redistribution doesn't violate the conservation of angular consequence.

Summary:

In CMGED, the conservation of angular momentum is not a passive law, but a consequence-managed behavior. The equation $L=I\cdot\omega$ holds true because:

- ω is kept constant by field-based regulation,
- I is allowed to evolve, but under controlled feedback,
- L is preserved globally, not locally, by system-wide ripple adjustments and mass-energy flux.

This explains:

- Why galactic rotation curves are flat,
- Why galaxies don't collapse,
- Why gravitational waves are necessary for balance,
- Why no external torque is needed — the system corrects from within.

5. Redshift as Rotational Frame Detuning

One of the most significant observational pillars of modern cosmology is the redshift of light from distant galaxies. Traditionally, redshift is interpreted through the lens of Doppler effect and metric expansion of space—meaning, the further a galaxy is, the faster it is receding from us, suggesting an expanding universe originating from a singularity (Big Bang).

However, this interpretation, though mathematically consistent, is not causally grounded in physical mechanics. It assumes recession without addressing the energetic or field-based basis for such motion. Moreover, it relies heavily on an ever-stretching, undefined fabric of space, leading to conceptual contradictions and an untestable notion of dark energy.

The Chavan Model of Galactic Energy Dynamics (CMGED) offers an alternative explanation: Redshift is not the result of expansion, but the consequence of rotational frame detuning—a mismatch in angular motion between different galaxies embedded in distinct segments of Structured Vacuum Energy (SVE).

5.1 Frame Mismatch Across Galaxies

In CMGED, each galaxy is a rotating annular structure (Ro–Ri) embedded within SVE, and the angular velocity (ω) of each galactic ring is determined by its internal energy distribution, mass configuration, and field interaction. Galaxies are not synchronized in rotation. Just as two spinning wheels may rotate at different speeds despite appearing still when viewed in isolation, galaxies too have their own rotational frames.

When light (or any energy signal) is emitted from a galaxy, it carries the frequency imprint of its local rotational time frame. If the observer's galaxy (such as ours) is spinning at a different angular speed, then the received frequency is detuned — experienced as a redshift (lower frequency) or blueshift (higher frequency) depending on the relative rotational alignment.

Redshift, in this model, is the natural result of comparing signals between two differently tuned galactic clocks.

This frame detuning is not hypothetical — it is measurable and field-anchored. It occurs not in "space" but in SVE domains, where rotation defines local time perception, and misalignment between two such domains leads to shifted signal interpretation.

5.2 Redshift Interpreted Through Rotational Frequency

Let us formalize this intuitively:

- Let Galaxy A rotate with angular velocity $\omega 1$, and Galaxy B with $\omega 2$.
- A photon emitted from Galaxy A is embedded with time-frequency structure based on $\omega 1$.
- Upon reception in Galaxy B (us), the photon is interpreted in our own time base $\omega 2$.

If $\omega 1 < \omega 2$, the received signal appears redshifted (lower frequency). If $\omega 1 > \omega 2$, the signal appears blueshifted (higher frequency).

Thus, the magnitude of redshift becomes a direct measure of rotational mismatch, not physical recession.

This perspective aligns elegantly with how resonance works in oscillatory systems: When two tuning forks of different frequencies interact, one perceives a beat or shift. The same happens here — but at the scale of galaxies and structured vacuum rings.

5.3 Why Redshift Increases with Distance: Not Expansion, But Divergence in Rotation

In the standard model, redshift increases with distance due to the assumed expansion of space. In CMGED, the same observational trend is explained by increasing rotational divergence as galaxies become more distant.

Why?

- Galaxies formed at different regions of the cosmic vacuum structure.
- Their initial angular velocities were set by localized SVE conditions, not a universal bang.
- As the distance between observer and source increases, so does the probability of encountering a galaxy with significantly different ω.
- Hence, redshift increases naturally with distance — not because they're moving away, but because they're spinning differently.

Furthermore, this model implies:

- Redshift is quantifiable in terms of angular speed ratios, not only recessional velocity.
- Non-linearity in redshift-distance relation is expected — some galaxies at similar distances may show different redshifts if their ω\omegaω differs significantly.

This can be tested empirically by:

- Measuring galactic rotation speeds independently (via HI line observations or Tully-Fisher relation),
- Comparing them with observed redshifts,
- Establishing correlations between ω mismatch and redshift, independent of distance.

If confirmed, this would disprove cosmic expansion as the dominant cause of redshift and validate CMGED's field-resonance explanation.

Summary:

Traditional View	CMGED Interpretation
Redshift = recession velocity due to expansion	Redshift = rotational frame detuning due to angular speed mismatch
Universe expanding uniformly	Galaxies spinning independently in structured SVE layers
Big Bang origin	Rotational emergence and consequence-regulated dynamics
Space is stretching	Perceived signal is misaligned due to time-frequency differential

Redshift, in this view, becomes not evidence of an origin event, but proof of a layered, rotating universe, where galaxies breathe, spin, and emit light according to their own rhythmic timeline.

6. Gravitational Waves as Internal Ripple Feedback

In mainstream physics, gravitational waves are described as oscillations in spacetime curvature, emitted when massive objects accelerate or collide, particularly during events like black hole mergers or neutron star collisions. These waves, predicted by general relativity and detected by instruments like LIGO and Virgo, are interpreted as temporary distortions in the fabric of space itself, rippling outward at the speed of light.

However, this concept—while mathematically elegant—lacks mechanical grounding. It presumes that space, a fundamentally dimensionless construct, can bend, stretch, and oscillate, without specifying what is actually "waving," what sustains it, or how the energy is regulated after emission.

In the Chavan Model of Galactic Energy Dynamics (CMGED), gravitational waves are understood fundamentally differently. They are not waves of geometry but structured pulses of realignment in the vacuum field — that is, they are ripples in Structured Vacuum Energy (SVE). These ripples serve a functional and intelligent purpose: to redistribute tension, rebalance angular momentum, and preserve rotational equilibrium across the galactic energy ring (Ro–Ri).

6.1 Not Spacetime Warps, But SVE Realignment Pulses

In CMGED, the universe is not a curvature-based construct but a field-based system. The vacuum is not passive — it is structured, intelligent, and responsive. Every mass-bearing, rotating, or radiating object exists within this structured medium and creates field tension as it interacts with its surroundings.

When a significant field imbalance occurs — for instance:

- A sudden mass inflow into the galactic core,
- A binary collapse event (e.g., black hole merger),
- A high-angular-velocity object decoupling from its orbital path,

— the vacuum field does not curve — it realigns.

This realignment emits a radial ripple through the SVE. These ripples, which we perceive as gravitational waves, are essentially field-level updates — a kind of vacuum resonance pulse propagating through the energy scaffold of the galaxy and beyond.

These are not distortions. They are corrections. Not geometric deformations, but consequence-driven energy harmonics.

6.2 Emitted from Galactic Centers or Extreme Collapses

The primary sources of such ripple feedback in CMGED include:

A. Galactic Cores

- The black hole at the center of a galaxy is not a gravitational well in curved space.
- Instead, it is a frictional sink of field tension, absorbing mass-energy that can no longer be structurally sustained in the outer disc.
- When too much energy enters the core too quickly, it triggers a vacuum tension spike, resulting in a pulse emitted radially

outward — a gravitational wave.
- This is the galaxy's exhale — its way of releasing stress and restoring balance.

B. Extreme Collapse Events

- When binary systems coalesce (e.g., neutron stars or stellar black holes), their combined angular momentum and mass density create a localized spike in field imbalance.
- The surrounding SVE structure reacts by realigning the rotational and mass-energy balance, sending out a ripple.
- These events often coincide with gamma-ray bursts or other transient signals, suggesting a multi-layered reconfiguration in vacuum and visible field.

In both cases, the waves are not byproducts of mass motion, but essential tools of field regulation.

6.3 Role in Rebalancing Angular Momentum Across Galactic Rings

A key insight of CMGED is that gravitational waves serve to redistribute angular momentum — not just transfer information.

Here's how:

- The galaxy maintains constant ω\omegaω (angular velocity) through mass-energy rebalancing.
- As mass migrates inward and moment of inertia I shifts, angular momentum $L=I\cdot\omega$ must be preserved.
- The field responds by emitting corrective ripples outward, subtly tweaking rotational speeds of stars and systems across the Ro–Ri band.

- These waves re-synchronize parts of the disc that may be drifting out of equilibrium due to age, local interaction, or star migration.

Think of it as a tuning fork reset — when one part goes out of tune, the ripple restores resonance across the structure.

Furthermore, these gravitational ripples act as non-destructive communication across the galactic structure, transferring angular information and enabling collective co-motion to continue.

This model not only explains the purpose of gravitational waves, but it also predicts:

- Ripple patterns originating from galactic cores, not just binary collapses.
- Low-amplitude, high-frequency background waves as part of ongoing galactic rebalancing — not just rare events.
- A possible correlation between spiral arm symmetry and internal ripple frequency.

Summary:

GR Interpretation	CMGED Interpretation
Gravitational waves are oscillations in space-time curvature	Gravitational waves are ripple corrections in the Structured Vacuum Energy
Emerge from mass acceleration and mergers	Emerge from mass-energy imbalance, core absorption, and dynamic field shifts
Propagate distortions in space	Propagate field tension realignments to rebalance angular momentum
Rare, high-energy signals	Regular, intelligent pulses regulating motion and structural order

In CMGED, gravitational waves are re-established not as exotic mathematical artifacts but as living pulses of the universe's attempt to stay in balance with itself.

Part III: Observational Correlations

7. Why Constellations Remain Fixed

One of the most intuitively powerful observations available to the human eye is the long-term stability of constellations. Over decades—indeed, centuries—constellations appear nearly identical in the night sky. Generations have looked up and seen Orion, Ursa Major, Cassiopeia, and others in the same familiar formations, season after season.

This apparent stillness has been historically attributed to the vast distances of stars and their very slow proper motion. While this explanation is mathematically defensible under classical mechanics, it lacks field-based causality and does not address why, despite massive galactic rotation speeds (e.g., the Sun orbiting the Milky Way at ~220 km/s), we observe no apparent drift among nearby stars within our galaxy.

The Chavan Model of Galactic Energy Dynamics (CMGED) offers a deeper, more causally sound interpretation: constellations appear fixed because their constituent stars are co-moving with us within the same rotational SVE ring. The Structured Vacuum Energy (SVE) governs not only the galactic dynamics but also what is seen as motion from the perspective of embedded observers.

7.1 Co-Motion in the Same SVE Ring

In CMGED, the stars in our sky—particularly those forming visible constellations—are not independent bodies flying through space at random velocities. Instead, they are nodes embedded within the same galactic energy ring (Ro–Ri) as our solar system.

This means:

- The Sun, Earth, and the surrounding stars are all rotating as part of a single dynamic energy structure, regulated by the field tension of the Milky Way's disc.
- Their positions relative to each other remain coherent over long timescales because they are riding the same field flow, much like leaves floating on the same river current.

The field itself is not inert; it is carrying all co-embedded systems with a shared angular speed ω\omegaω. Hence, there is no "differential motion" within the local SVE frame to cause noticeable drift of star positions from our viewpoint.

In this model, constellations do not hold still because motion is absent. They remain fixed because motion is shared.

7.2 Proper Motion vs. Structural Stillness

Modern astronomy accounts for proper motion—the actual motion of stars relative to the Sun—by tracking their changing positions over time with precise instruments (e.g., Gaia mission). But the changes are minuscule, typically fractions of an arcsecond per year. This raises a deeper question:

Why is proper motion so small when galactic speeds are so large?

CMGED explains this elegantly: Proper motion becomes visible only when field-level resonance begins to drift. That is:

- A star may exhibit a measurable motion only when it starts decoupling from the rotational field equilibrium.
- This might occur due to age, mass loss, gravitational interactions, or entry into a denser or more diffuse field region.

But as long as the star is energetically bound to the same SVE ring, its motion relative to the observer is harmonized, not independent.

Thus, what is classically perceived as "almost no motion" is, in fact, perfectly coherent motion within the same galactic resonance frame.

7.3 Motion Becomes Visible Only When the Background Is Misaligned

CMGED introduces a critical relational principle:

Motion is not an absolute quantity—it is a consequence of frame mismatch.

In observational terms:

- We perceive motion only when we and the object we are observing are embedded in different field frames.
- Within the same field structure (e.g., the Ro–Ri ring of the Milky Way), objects maintain relative stillness.
- But when we observe objects outside our field (e.g., distant galaxies), their motion becomes visibly expressed—through redshift, blueshift, or rotational alignment.

This is why:

- Planets are seen to move across the sky — because they orbit within a local sub-structure (the solar system) that includes internal differential speeds.
- Stars appear fixed — because they share our galactic motion.
- Distant galaxies show redshift — because they spin in different SVE frames with mismatched angular velocities.

This redefines relativity itself not in abstract terms of spacetime frames, but in field-tension harmonics:

When the background resonates with you, motion is hidden. When it detunes, motion becomes visible.

Summary:

Observation	Classical Interpretation	CMGED Interpretation
Constellations appear fixed	Stars are too far and move slowly	Stars share co-motion in the same SVE ring
Proper motion is small	Long distances mute real motion	Motion is masked by field-level resonance
Motion becomes visible at large scales	Due to faster motion of galaxies	Due to rotational detuning across SVE fields

In CMGED, the sky is not still — it is perfectly synchronized. The stars do not lack motion; they are simply moving together, with us, on the same silent wheel.

8. Galactic Mergers as Field Rebalancing Events

In conventional astrophysics, galactic mergers are described as gravitational interactions between two galaxies, leading to eventual coalescence through a process of chaotic collision, tidal distortion, and violent redistribution of stars and gas. These events are treated as rare but inevitable consequences of a dynamic, expanding universe where galaxies drift randomly and gravitationally attract each other over time.

However, this view is incomplete and inconsistent with the broader order observed in galactic structures. The chaotic nature attributed to mergers stands in stark contrast to the symmetry, balance, and persistence seen in galaxies that survive for billions of years. More importantly, it fails to account for what triggers such interactions and why they follow certain predictable alignments and patterns.

In Chavan's Model of Galactic Energy Dynamics (CMGED), galactic mergers are not random gravitational accidents. They are systemic field-level rebalancing events, occurring only when a galaxy fails to maintain angular equilibrium within its Structured Vacuum Energy (SVE) field. A merger, in this context, is not a crash — it is a consequence-driven reset, a higher-order field response to an imbalance that can no longer be corrected internally.

8.1 Not Chaotic Collisions

CMGED challenges the idea that galaxies "collide" in the conventional sense. The vast distances between stars make direct collisions rare even in merging systems. What is observed are interlocking field dynamics—gravitational stretching, tidal bridges, gas compression, and starburst phenomena—all of which are secondary effects of field convergence, not impact.

In this model:

- Two galaxies begin to interact not because they drifted too close, but because one or both have entered a phase of field instability.
- This instability arises from failure to conserve angular momentum through internal mass-energy redistribution (as explained in Section 4).
- When such imbalance exceeds the regulatory capacity of ripple feedback and local rebalancing, the galaxy becomes susceptible to merging with a nearby system that can absorb or redistribute the excess tension.

Thus, a merger is not chaos—it is vacuum logic seeking balance.

8.2 Consequence-Triggered Resets When Angular Balance Fails

Every galaxy in CMGED exists as a rotationally regulated ring within the SVE field. As long as it maintains its:

- Constant angular speed ω,
- Field-aligned moment of inertia I,
- Ripple-regulated angular momentum L,

—it remains self-sustaining and stable.

But when the internal energy distribution becomes too lopsided—due to:

- Excessive mass inflow toward the core,

- Starburst-induced mass ejections,
- Long-term asymmetry in spiral arms,
- Loss of ripple feedback due to energetic damping—

the galaxy fails to preserve its angular consequence. At this stage, it begins to lose coherence as a standalone system.

That failure does not lead to collapse, but to integration. The galaxy is now drawn, not by Newtonian gravity alone, but by field-level attraction toward a compatible SVE node — often a neighboring galaxy whose field geometry can absorb the imbalance and initiate a new equilibrium structure.

A merger is thus a reset in rotational logic, not a destructive event. It is the galaxy's way of resolving its imbalance by becoming part of a new structure.

8.3 Part of a Higher-Order Rotational Field Dynamic

Perhaps most profoundly, CMGED asserts that galactic mergers are not isolated incidents. They are embedded within a nested hierarchy of rotational field dynamics.

Just as:

- Planets orbit stars within SVE-regulated solar systems,
- Stars orbit galactic centers within Ro–Ri bands,
- Galaxies themselves orbit within supercluster-scale SVE structures.

Thus, a merger is often not just a two-body field convergence — it is a synchronized phase correction within a larger rotating scaffold.

This explains:

- Why many mergers are aligned along filaments and sheets of the cosmic web.
- Why galaxies in the same region of the universe exhibit spin vector coherence.
- Why merger remnants often exhibit new, well-organized rotational behavior (e.g., elliptical galaxies with conserved total angular momentum).

In CMGED, no motion is random, and no interaction is uncaused. Mergers are part of the cosmic orchestration of balance, dictated not by mechanical impact but by energetic necessity and field alignment.

Summary:

Conventional View	CMGED Interpretation
Mergers are random gravitational collisions	Mergers are field-regulated resets due to angular imbalance
Chaos dominates the process	Vacuum field seeks new equilibrium through integration
Galaxies collide due to proximity	Galaxies merge due to systemic field-level necessity
Mergers disrupt order	Mergers re-establish higher-order order

Mergers, in CMGED, are not catastrophic—they are constructive. They are the vacuum field's way of saying: "This structure must now evolve."

9. Distant Galaxies and Time Perception

In conventional cosmology, time is treated as a universal coordinate—a fourth dimension flowing uniformly across space, modulated only by relative velocity or gravitational potential (as described in Einstein's relativity). Time, in this framework, is synchronized across the cosmos through a shared history of expansion, rooted in the Big Bang.

However, this concept is deeply flawed. It assumes that all galaxies experience the flow of time identically, regardless of their internal energy structure, rotational dynamics, or position within the vacuum field. This assumption collapses under scrutiny, especially in the face of mounting evidence for:

- Variable rates of cosmic evolution,
- Asymmetrical time dilation in high-energy systems,
- Observed redshift phenomena that do not always correlate cleanly with distance.

Chavan's Model of Galactic Energy Dynamics (CMGED) introduces a radical and necessary revision: time is not universal—it is rotationally anchored. Each galaxy, as a dynamically rotating system embedded in Structured Vacuum Energy (SVE), operates on its own angular time base. That is, the very perception of time is local to the rotational rhythm of its galactic ring.

9.1 Different Galaxies → Different Angular Time Bases

In CMGED, galaxies are not passive containers of stars drifting in expanding space. They are annular rotational systems (Ro–Ri rings), where angular velocity ω\omegaω governs not just motion, but the emergent perception of time itself.

If we accept that:

- Planetary systems have different orbital periods (e.g., Earth: 365 days; Neptune: 165 years),
- Galaxies too must have distinct rotational periods and field-defined harmonic frequencies,
- Then each galaxy exists within a different rhythm of energetic change, and thus a different clock.

A galaxy's time is not ticking on our wall clock — it is beating with its own rotational heart.

The implication is profound: Distant galaxies live in different temporal realities. Not because of relativistic distortion, but because their internal angular structure defines a unique SVE time base — a tempo of change governed by consequence.

9.2 Why Cosmic "Time" Is Not Universal

In this framework, "cosmic time" — the idea of a single timeline that applies uniformly to all observers — is an illusion born of oversimplified models.

Here's why:

- Time arises only through change, and change is structured through rotation, spin, and mass-energy reconfiguration.
- Since each galaxy has a different configuration of mass, angular momentum, and field alignment, the rate of change, and therefore the perception of time, varies across systems.
- Our measurement of cosmic events from Earth assumes that

our own field-defined time base applies to distant systems — but this is projection, not fact.

For example:

- A supernova in a distant galaxy may appear to occur over weeks in our observation frame,
- But in its own time base, the process may be drastically different — either stretched or compressed,
- We interpret it through our galactic rhythm, misaligning the true temporal behavior of that system.

Hence, there is no absolute "now" in the cosmos. There are only localized rotational timelines shaped by each galaxy's embedded dynamics.

9.3 Redshift as a Signature of Layered Rotational Timelines

Redshift is classically interpreted as Doppler stretching—light from distant galaxies appears redder because they are "moving away" in an expanding universe.

In CMGED, redshift takes on a deeper meaning: It is a temporal echo of angular mismatch.

If each galaxy has its own rotational time base:

- A photon emitted from a galaxy rotating at angular velocity $\omega 1$ will carry the frequency imprint of that galaxy's time base.
- When received by a galaxy rotating at $\omega 2$, the signal is interpreted relative to its own time base.
- If $\omega 1<\omega 2$, the signal appears redshifted — a slower beat arriving in a faster rhythm system.

- This redshift is not a speed effect; it is a temporal detuning — a mismatch in rotational clocks.

This reveals that redshift is not only spatially layered but also temporally layered: The further we look, the more temporal misalignment we observe, Because distant galaxies are increasingly likely to be out of sync with our time frame, Redshift, then, is a diagnostic tool of cosmic time variation, not expansion.

The universe is not flying apart. We are simply hearing other galaxies' rhythms through the distortion of our own beat.

Summary:

Concept	Standard View	CMGED Perspective
Time	Universal, linear, global	Local, rotational, field-defined
Redshift	Velocity-based stretching	Rotational frame detuning
Cosmic evolution	Uniform timeline	Layered, system-specific timelines
Now	Shared across the universe	Unique to each rotating system

In CMGED, time is not an arrow fired from a mythical singularity. It is a wheel—and each galaxy spins its own.

Part IV: Theoretical Implications

10. Collapse as Equilibrium, Not Destruction

In prevailing astrophysical narratives, the term "collapse" evokes images of violent ends — of stars exploding in supernovae, of black holes tearing apart matter, of galaxies falling into disorder, chaos, and gravitational oblivion. The death of celestial bodies is often seen as final, dramatic, and inherently destructive.

However, this perspective stems from a limited, force-based worldview, where gravity is the ultimate ruler, and energy serves no higher principle than attraction. It ignores the subtler, structured nature of the universe — the intelligent organization evident in rotational stability, recurrence, and harmony.

In Chavan's Model of Galactic Energy Dynamics (CMGED), collapse is redefined not as an ending, but as a transition into equilibrium. It is a field-governed return to balance, not a breakdown. The core premise is simple yet profound:

What appears as destruction is often a dissolution — a graceful return of structured energy into the foundational medium, the Structured Vacuum Energy (SVE).

This reinterpretation fundamentally shifts how we understand black holes, galactic deaths, and the larger process of cosmic evolution.

10.1 Black Hole as a Frictional Sink

In mainstream cosmology, black holes are treated as mysterious, destructive points of infinite density — singularities where known physics breaks down. They are often viewed as cosmic anomalies that "swallow" everything, bending space and time into nothingness.

In CMGED, the black hole at a galactic center is re-envisioned not as a singularity, but as a frictional sink within the SVE field — a natural endpoint of regulated mass-energy flow, not a breakdown of reality.

Here's how:

- As stars evolve and mass migrates inward across the galactic ring (Ro–Ri), the central region accumulates tension.
- This inward migration is not collapse due to gravitational pull, but a field consequence — a slow rebalancing of angular momentum.
- The black hole acts as the point of tension absorption, where excess energy is converted into field realignment, not annihilation.

Thus, the black hole is a cosmic balancer, not a destroyer. It is a silent coordinator, drawing in that which no longer maintains structural motion — turning localized imbalance into universal stillness.

10.2 Dissolution of Galaxies into SVE

Galaxies, like all systems in CMGED, are finite energy structures embedded in a larger field. They form, rotate, breathe, and eventually — when their internal mechanisms can no longer sustain rotational coherence — dissolve back into the SVE field.

This process is not catastrophic. It is akin to:

- A spinning top slowing down and eventually stopping,
- A wave collapsing back into the ocean that gave it form.

Galactic dissolution follows a clear sequence:

1. Mass-energy migration inward gradually increases the field tension at the core.
2. Ripple feedback (gravitational waves) becomes weaker over time as the structure loses coherence.
3. The system can no longer preserve its angular momentum through redistribution.
4. Stars fall inward, spiral arms fragment, and the ring's Ro–Ri structure dissolves.
5. All energy is absorbed, neutralized, and returned to the structured vacuum — without explosion, without entropy-driven chaos.

The galaxy does not end. It rejoins the breath of the universe. This is recycling, not death. The energy is not lost — it is de-patterned, freed to reform in another expression, perhaps in another region of the field.

10.3 Cosmic Recycling Without Chaos

The universe, as seen through CMGED, is not a battleground of forces — it is a self-balancing continuum of structured energy interactions, where collapse is simply re-integration.

Consider:

- A collapsing star becomes a black hole, which becomes a regulator.
- A dissolving galaxy returns its energy to the vacuum, which seeds new structures elsewhere.
- Mergers are not accidents, but orchestrated redistributions.
- Gravitational waves are not tremors of destruction, but ripples of renewal.

This model reveals a universe that is not expanding into disorder, but constantly recycling through localized order and global balance.

There is no final collapse. Only completion — followed by continuation.

Summary:

Traditional View	CMGED Perspective
Collapse = destruction	Collapse = return to equilibrium
Black holes = singularities	Black holes = frictional sinks for excess energy
Galactic death = gravitational collapse	Galactic dissolution = graceful reintegration into SVE
Universe trends toward entropy	Universe maintains dynamic balance through regulated recycling

In this view, the universe does not end in a Big Freeze or Big Rip. It breathes — in rings and ripples, endlessly folding energy back into the vacuum, preparing it for new cycles, new harmonics, and new systems.

Collapse is not a tragedy. It is the field's most intelligent act of letting go.

11. Nested Rotational Universe Model

The more we observe the universe, the more it reveals a striking consistency: everything rotates. From the spin of subatomic particles to the orbit of planets, the rotation of stars, the co-motion of galaxies, and even the slow swirl of galaxy clusters — the signature of spin pervades all cosmic structures. Yet this rotation is not chaotic. It is not disordered. It is nested, layered, and harmonized — governed by field-level balance and a universal search for equilibrium.

Chavan's Model of Galactic Energy Dynamics (CMGED) recognizes this not as coincidence, but as cosmic architecture. The universe is built on rotational symmetry, with systems embedded inside larger systems, each spinning not to accelerate into disorder, but to stabilize, regulate, and breathe within a structured energetic medium — the Structured Vacuum Energy (SVE) field.

This vision culminates in the Nested Rotational Universe Model, where spin is not a byproduct of formation, but a requirement of continued existence.

11.1 From Planets to Stars to Galaxies to Clusters to SVE Rings

The Nested Rotational Model describes the universe as a hierarchy of co-rotating systems, each embedded in a larger energy framework, governed by the same principles of angular momentum, field balance, and ripple feedback.

Let us examine this cosmic layering:

1. Planetary Systems (Core Mechanics of PED)

- Planets orbit stars within Ro–Ri bands defined by the star's local vacuum field.
- Their angular speeds vary but remain regulated by the field tension and spin of the parent star.
- Spins and tilts emerge as responses to internal mass distribution and external field harmonics.

2. Stellar Orbits in Galaxies (CMGED Layer 1)

- Stars, along with their planetary systems, orbit galactic cores as part of a shared galactic energy ring.
- These stars are not individually free but co-moving nodes embedded in a field-regulated Ro–Ri annulus.
- Spiral arms form as regions of coherent rotational rhythm.

3. Galaxies in Clusters and Filaments (CMGED Layer 2)

- Galaxies themselves are not stationary. They orbit or drift within super-galactic fields, maintaining orientations and spin axes often aligned across millions of light-years.
- These fields may form giant rotating filaments or ring structures, suggesting higher-order Ro–Ri patterns.
- Mergers and interactions occur as part of field-level realignments.

4. SVE Rings at Universal Scale (CUED: Cosmic UED Layer)

- All these nested layers exist within the grand structure of SVE rings — vast, slowly rotating tension bands within the structured vacuum field itself.

- These define the "zones of coherence" across the observable universe and are responsible for:
    - Aligning spin axes of galaxy clusters,
    - Establishing regional field frequencies,
    - Governing the broader rhythm of cosmic time and motion.

The universe is not a chaotic expanse — it is a nested gyroscope, where every system spins to remain where it is.

11.2 Everything Spins to Stay Still — and Breathes to Stay Balanced

One of the most poetic truths revealed by CMGED is that spin is not motion for the sake of motion. It is the mechanism by which a system remains in place.

- Planets do not spiral into their stars because their spin and revolution lock them into dynamic balance.
- Stars remain stable in galactic arms not through inertia, but because they are continually adjusted through ripple feedback and field tension.
- Galaxies orbit in clusters not because they are flung there, but because they are sustained there by rotational tuning and nested field coherence.

This leads to a powerful realization:

Stillness is an illusion created by perfect motion. Balance is maintained not by force, but by structured spin. And the universe breathes — not with lungs, but with rings. This breath is not metaphorical. It is mechanical and measurable:

- Mass flows inward toward centers (inhale),

- Ripple waves restore balance outward (exhale),
- Collapse returns energy to the field (pause),
- And new structures emerge (restart the cycle).

Each system — from atom to galaxy — is part of this grand breath, anchored by spin and regulated by consequence.

Summary:

Scale	System	Rotation Role	Balance Mechanism
Planetary	Planets around stars	Orbital anchoring	Local SVE field tension
Galactic	Stars around cores	Disc co-motion	Ripple feedback & angular regulation
Supergalactic	Galaxies in clusters	Orientation coherence	Filamentary spin alignment
Cosmic	SVE rings	Rotational scaffolding	Tension harmonics across nested fields

In this model, the universe is alive with organized motion. It is not random. It is not expanding into meaninglessness. It is rotating into structure, dissolving into balance, and re-emerging through consequence. The nested rotational universe is not theory. It is the natural language of balance, speaking through rings, ripples, and the rhythm of intelligent motion.

12. Conclusion: The Chavan Framework for Galactic Order

The universe is not expanding into silence, nor collapsing into chaos. It is rotating into balance — system by system, ring by ring, ripple by ripple. What we observe across cosmic scales — from the silent fixity of stars to the spiraling elegance of galaxies — is not randomness held together by gravity, but intelligence structured into motion, embedded in a living medium called Structured Vacuum Energy (SVE).

Chavan's Model of Galactic Energy Dynamics (CMGED) is a response to the limitations and paradoxes of mainstream cosmology. It is not merely an alternative theory — it is a paradigm shift, grounded in physical logic, mechanical causality, and observable consistency. It does not reject classical physics or general relativity — it contextualizes them, restoring meaning to motion, coherence to collapse, and balance to a universe that has too long been explained through abstraction, patches, and paradoxes.

12.1 What CMGED Replaces

Let us be precise. CMGED replaces:

- The fictional scaffolding of dark matter with the real structure of vacuum energy tension.
- The distorting lens of spacetime curvature with rotational equilibrium in nested energy fields.
- The misinterpretation of redshift as recession with rotational frame detuning across SVE domains.
- The catastrophic view of black holes and collapse with the understanding of frictional sinks and cosmic recycling.

- The illusion of universal time with rotational time bases unique to each system.

12.2 What CMGED Establishes

This framework lays the foundation for a new cosmological logic:

- Galaxies are not collections of mass orbiting freely but are structured energy rings (Ro–Ri) embedded in SVE.
- Motion is not inertial — it is a consequence of field imbalance, and spin is how systems hold equilibrium.
- Gravitational waves are not spacetime tremors but ripple feedback mechanisms for realignment.
- Constellations remain fixed because of co-motion within the same field ring, not due to distance alone.
- Mergers are not accidents — they are restructuring events when angular regulation fails.
- Time is not global — it is system-specific, governed by angular velocity and field resonance.

12.3 What CMGED Unlocks

By integrating planetary dynamics, galactic architecture, rotational feedback, and vacuum structuring, CMGED opens the door to:

- Predictive cosmology without reliance on hypothetical matter or energy.
- A mechanically testable framework for understanding redshift, galaxy formation, and gravitational phenomena.
- A complete alternative to the Big Bang, rooted in structure, not explosion.
- The expansion of UED (Universal Energy Dynamics) to encompass cosmic and sub-cosmic scales through field-regulated consequence.

12.4 The Cosmic Message Behind the Framework

More than equations or theories, CMGED tells us something deeper: The universe is not empty. Not violent. Not accidental.

It is:

- Spinning to stay whole
- Breathing to stay in balance
- Rippling to remember its shape
- Collapsing not to end — but to return

There is no waste, no randomness. Only nested rings, field logic, and the law of consequence acting through time-shaped motion.

Final Statement:

The Chavan Framework for Galactic Order restores what cosmology lost: coherence. It does not replace wonder with certainty. It brings wonder into structure. And it teaches us that to understand the stars, we must first learn to see how everything moves — together — in silence, not noise.*

This is not the end of cosmology. It is the return of balance.

Appendices

Appendix A

Comparative Table: CMGED vs Mainstream Models

This appendix presents a side-by-side comparison between mainstream cosmological frameworks and the Chavan Model of Galactic Energy Dynamics (CMGED) across key areas of galactic structure, motion, perception, and cosmological interpretation. The objective is to highlight not only the conceptual differences, but also the logical coherence, explanatory power, and observational compatibility of CMGED.

A1. Galactic Rotation and Structure

Aspect	Mainstream Model	CMGED
Galactic Rotation Curves	Flat rotation curves explained by hypothetical dark matter halos	Explained by dynamic mass-energy redistribution within a rotating Ro–Ri annular structure governed by SVE
Nature of Galaxy	A collection of gravitationally bound stars orbiting a central mass	A rotating structured field system with co-motion across embedded nodes (stars, systems)
Stability Mechanism	Gravitational equilibrium + dark matter	Angular momentum regulation via ripple feedback and field tension

A2. Redshift and Cosmic Motion

Aspect	Mainstream Model	CMGED
Redshift Cause	Recession velocity from cosmic expansion (Doppler or metric)	Rotational frame detuning — mismatch in angular frequency between emitter and observer galaxy
Interpretation of Distance	More redshift = farther + faster receding	More redshift = greater rotational divergence across SVE layers
Cosmic Expansion	Space itself is stretching	No expansion — galaxies exist in rotational equilibrium within nested vacuum fields

A3. Gravity and Black Holes

Aspect	Mainstream Model	CMGED
Nature of Gravity	Curvature of space-time by mass (GR)	Tension in structured vacuum energy (SVE) regulating motion by consequence
Black Holes	Singularities that collapse matter into infinite density	Frictional sinks that absorb and neutralize excess field tension — equilibrium regulators, not destroyers
Collapse	Violent, chaotic, entropic end-state	Controlled dissolution into SVE — a phase of return and recycling

A4. Time and Perception

Aspect	Mainstream Model	CMGED
Nature of Time	Global time modulated by gravity or velocity (relativity)	Local, rotational time bases — time emerges from each system's angular dynamics within its field
Uniform Cosmic Clock	Assumed across all space-time	Replaced by layered field-specific timelines
Redshift-Time Link	Higher redshift = older in universal time	Redshift = signature of different temporal rhythm, not linear age

A5. Gravitational Waves and Field Communication

Aspect	Mainstream Model	CMGED
Gravitational Waves	Oscillations in space-time curvature from acceleration of mass	Ripple feedback mechanisms in SVE — field realignment pulses to preserve system-wide balance
Source	Binary mergers, neutron stars, black hole collisions	Core tension spikes, collapse events, and systemic rebalancing in Ro–Ri structures
Purpose	Observational signature of violent events	Functional regulation of angular momentum and structural harmony

A6. Galactic Mergers and Evolution

Aspect	Mainstream Model	CMGED
Mergers	Random gravitational collisions between drifting galaxies	Consequence-triggered resets — field-level events when rotational equilibrium fails
Outcome of Mergers	Chaotic redistribution, starbursts, loss of order	Integration into higher-order field dynamics — new equilibrium formation
Cosmic Evolution	From initial chaos to increasing structure (entropy-dominated)	From localized order → imbalance → correction → reintegration — a breathing field

A7. Philosophical Implications

Aspect	Mainstream Model	CMGED
Origin of the Universe	Big Bang — singular explosive event	Structured emergence through nested rotational harmonics — no singularity required
Ultimate Fate	Heat death, Big Rip, or collapse	Perpetual recycling via field rebalancing and rotational transitions
Nature of Reality	Mechanistic, geometry-driven, force-based	Consequential, structured, intelligence-bound system regulated by field balance

Summary View:

Domain	Mainstream View	CMGED View
Core Medium	Empty space or space-time fabric	Structured Vacuum Energy (SVE)
Motion	Force-based or inertial	Consequence-based and regulated by field feedback
Collapse	Destructive, irreversible	Recycling, field-dissolution, and equilibrium
Stillness	Inert or absent motion	Co-motion within same field frame
Expansion	Fundamental assumption	Field misinterpretation — illusion of divergence

This comparative model illustrates that CMGED is not merely compatible with observation — it explains it more fundamentally. It removes the need for hypothetical constructs and replaces them with logical, field-consistent mechanisms grounded in rotational structure and consequence.

Appendix B

Observational Data Points Supporting CMGED

(Gaia, SDSS, LIGO and Other Empirical Datasets)

Chavan's Model of Galactic Energy Dynamics (CMGED) is rooted in physical logic and field-based mechanics but draws its strength from its alignment with observational reality. While the model proposes a conceptual shift from conventional interpretations, it does not reject observational data—instead, it reinterprets that data through a more coherent framework of structured energy dynamics and rotational consequence.

This appendix outlines major datasets from leading astronomical observatories and missions (such as Gaia, SDSS, and LIGO), showing how their observations:

1. Validate CMGED's key predictions,
2. Expose cracks in mainstream models,
3. Support a field-regulated, rotationally structured cosmology.

B.1 Gaia Mission (ESA) – Precision Astrometry and Proper Motion

The Gaia space observatory, launched by the European Space Agency, is mapping over 1.8 billion stars with unprecedented accuracy, recording their positions, proper motions, parallaxes, and velocities.

Key Observations:

- Stars in the Milky Way exhibit coherent large-scale motion, forming clear kinematic structures.

- Proper motions of stars are surprisingly small across constellations, despite the galaxy's high rotational speed.
- Gaia data reveals "moving groups" — collections of stars with shared trajectories, embedded in spiral arms.

CMGED Interpretation:

- Stars do not move freely in empty space; they are co-moving within the same galactic energy ring (Ro–Ri).
- Proper motion is minimized due to field resonance — motion becomes visible only when a star begins to detune from the field.
- The "moving groups" are not drifting relics, but field-anchored harmonics within the rotational shell of the galaxy.

Gaia supports CMGED's premise that motion is not observed when field alignment is preserved, reinforcing the concept of perceived stillness via co-motion.

B.2 SDSS (Sloan Digital Sky Survey) – Galaxy Maps and Redshift Surveys

The Sloan Digital Sky Survey (SDSS) has created 3D maps of millions of galaxies, quasars, and stars. It has provided redshift data, galaxy morphology, spin alignment, and cluster distributions across large cosmological volumes.

Key Observations:

- Redshift increases with distance, but non-uniformly across galactic types and directions.
- Some galaxies with similar distances show significantly

different redshifts.

- Spin alignments of galaxies are observed in certain large-scale structures (e.g., filaments and walls).
- Galaxy clusters often show coherent motion and structure across megaparsec scales.

CMGED Interpretation:

- Redshift is not purely velocity-based, but a signature of rotational time base mismatch between galaxies.
- Differences in redshift at similar distances support CMGED's prediction of non-expansion-based redshift variations.
- Spin alignment across vast regions supports the idea of nested rotational SVE rings, where galaxies are embedded in higher-order co-rotating fields.
- Coherent cluster motion aligns with CMGED's claim that galactic systems are not randomly placed, but field-bound and phase-aligned.

SDSS thus provides strong empirical ground for reinterpreting redshift as rotational detuning and supports the concept of a field-layered, non-expanding universe.

B.3 LIGO-Virgo – Gravitational Wave Detections

The LIGO (Laser Interferometer Gravitational-Wave Observatory) and its European counterpart Virgo have detected gravitational waves from high-energy events like binary black hole mergers and neutron star collisions.

Key Observations:

- Detected gravitational waves are consistent with models of

massive, compact object mergers.
- The waveforms are often symmetric, radial, and transient, lasting fractions of a second.
- These events are extremely rare compared to the expected number of high-energy events.

CMGED Interpretation:

- These are ripple corrections in the SVE field, not spacetime warps.
- The radial symmetry and pulse-like nature match CMGED's prediction of field realignment waves—ripples generated when internal field tension is rapidly redistributed.
- Their rarity suggests that such ripple events are regulated equilibrium responses, not constant radiation — aligning with CMGED's model of gravitational waves as energy-balancing events, not random or frequent disturbances.

Additionally, CMGED predicts:

- Weaker, continuous gravitational background signals as part of galactic breathing,
- Ripple precursors prior to major collapse events (yet undetected in mainstream),
- And a broader spectrum of gravitational phenomena that LIGO-class instruments may detect in the future, if tuned for structured field feedback rather than Einsteinian curves.

B.4 Additional Supporting Observations (General)

Dataset	Observation	CMGED Insight
Hubble Deep Field	Galaxies are distributed in layers and aligned filaments	Supports nested rotational structure, not random expansion
Cosmic Microwave Background (CMB) Dipole	Suggests solar system motion relative to CMB	Consistent with SVE ring co-motion and a field-anchored cosmic frame
Galaxy Spin Surveys (e.g., Galaxy Zoo)	Non-random distribution of galaxy spin directions	Evidence for coherence in higher-order SVE field layers
Supernova Light Curves	Time-dilation not always uniform with redshift	Supports CMGED's idea of local time bases affecting observed decay rates

Summary:

Mission	Conventional Use	CMGED Contribution
Gaia	Stellar mapping and motion	Validates co-motion, ripple-based equilibrium, minimal relative drift
SDSS	Redshift catalog and galaxy distribution	Confirms rotational redshift and large-scale spin alignment
LIGO-Virgo	Spacetime wave detection	Reinterpreted as field ripple correction mechanisms
Hubble, CMB, Galaxy Zoo	Cosmological surveys	Support for nested, aligned, and rotationally governed structure

Together, these observational programs provide a multi-scale validation of CMGED — not by invention of new phenomena, but by correctly interpreting existing ones within a unified, field-based model of consequence and rotation.

Appendix C

Glossary of UED and CMGED Terms

This glossary defines the core terms and concepts used throughout the frameworks of Universal Energy Dynamics (UED), Core Mechanics of Planetary Energy Dynamics (CMPED), and the Chavan Model of Galactic Energy Dynamics (CMGED). These terms reflect a paradigm shift from traditional force-based cosmology to a consequence-driven, field-regulated understanding of the universe.

Structured Vacuum Energy (SVE)

Definition: The foundational medium of the universe, composed not of empty space but of structured, latent energy capable of storing tension, redistributing motion, and governing all dynamics via equilibrium-seeking interactions.

Role:

- Acts as the "field" within which all physical systems are embedded.
- Responsible for the emergence of motion, structure, and feedback through ripple mechanics.
- Replaces the abstract concept of spacetime with a real, energetic, dynamic continuum.

Ro–Ri Structure

Definition: The annular region within which a rotational system (planetary system, galaxy, etc.) exists and maintains balance.

- Ro (Outer Radius): Boundary beyond which field coherence

weakens.

- Ri (Inner Radius): Threshold near the core where frictional tension begins rising steeply.

Role:

- Describes the dynamic zone of equilibrium for rotating systems.
- Governs structural stability through internal mass-energy distribution.

Ripple Feedback

Definition: Radial energy pulses emitted through SVE when a system undergoes structural tension changes. These are perceived as gravitational waves, but functionally serve as field-corrective mechanisms.

Role:

- Balances shifts in angular momentum.
- Redistributes tension across the rotating shell.
- Prevents collapse by maintaining system-wide coherence.

Consequence-Driven Motion

Definition: Motion that arises not from arbitrary force application, but from field imbalance, mass-energy asymmetry, or rotational misalignment that necessitates a shift toward equilibrium.

Role:

- Replaces Newtonian inertia with a more intelligent model.
- Applies at all scales: atomic, planetary, galactic, cosmic.
- Establishes motion as evidence of field tension, not freedom.

Angular Time Base

Definition: The effective rate of time progression defined by a system's rotational velocity within its SVE field. Each system's time is emergent, not absolute.

Role:

- Replaces the concept of "universal time."
- Explains why redshift varies with field resonance and why temporal measurements between galaxies cannot be standardized.

Rotational Detuning

Definition: The misalignment in angular frequency between two rotating systems (e.g., galaxies) that leads to redshift or blueshift when signals are exchanged.

Role:

- Redefines redshift as a temporal and rotational mismatch, not recession velocity.
- Enables layered interpretation of cosmic structure and motion.

Field Resonance

Definition: The energetic alignment or harmony between a system and its surrounding SVE structure. When resonance is maintained, systems remain perceptually still. When broken, motion becomes observable.

Role:

- Governs how and when motion is perceived.

- Central to the concept of why constellations remain fixed.
- Explains nested stillness through shared motion.

Frictional Sink

Definition: A high-tension region within SVE (e.g., a black hole) where field imbalance is neutralized. Contrary to being destructive, these sinks absorb excess angular tension, aiding in energy recycling.

Role:

- Stabilizes systems approaching collapse.
- Serves as a terminal node for mass-energy return to the field.
- Regulates the end-phase of galactic dynamics without entropy violation.

Nested Rotational Universe

Definition: A cosmic structure where systems at every scale — from particles to galaxies to clusters — are co-rotating within larger systems, each governed by field-regulated angular logic.

Role:

- Describes the deep symmetry and hierarchy of spin across the universe.
- Explains coherence of motion from local to cosmic scale.
- Foundation for non-expanding, consequence-based cosmology.

Dynamic Equilibrium

Definition: A continuously self-correcting state in which energy, mass, and motion are regulated through feedback mechanisms in the field.

Role:

- Prevents collapse or dissipation.
- Maintains the stability of Ro–Ri zones.
- Core principle of all structure persistence in UED.

Collapse as Equilibrium

Definition: The process by which systems return their energy to the vacuum field not through chaos, but as a structured reintegration when rotational regulation ends.

Role:

- Replaces entropy-driven destruction with organized dissolution.
- Key to understanding black holes, mergers, and galactic death.
- Harmonizes cosmic life cycles with field balance.

Galactic Breathing

Definition: The periodic inward migration of mass-energy and outward emission of ripple feedback that keeps a galaxy in rotational harmony.

Role:

- Maintains system-wide regulation of I, ω, and L.
- Describes how galaxies live, not just rotate.
- Symbol of the living nature of the field.

Law of Consequence (LOC)

Definition: The fundamental governing principle in UED: Every action within a field generates a response aimed at restoring balance, not by force, but through structured feedback.

Role:

- Replaces Newton's third law and spacetime curvature simultaneously.
- Operates across energy scales.
- Gives cosmology a moral symmetry — structure arises only from sustained consequence.

This glossary not only defines technical terms but reinforces the central philosophy of CMGED: The universe is not a place of objects governed by laws—it is a structured field where everything exists because of what everything else does.

Appendix D

Proposed Experiments and Observational Tests

While the Chavan Model of Galactic Energy Dynamics (CMGED) is deeply grounded in observational coherence and theoretical clarity, its real strength lies in its ability to generate testable predictions and falsifiable hypotheses. This is what separates a philosophical framework from a scientific one.

This appendix outlines a series of practical observational strategies and experiments, leveraging current or near-future capabilities, to test the predictions of CMGED. These proposals focus on redshift reinterpretation, gravitational wave dynamics, time base divergence, and rotational field harmonics.

D.1 Rotational Redshift Mapping

Objective: To validate that redshift is caused by rotational frame detuning, not cosmic expansion.

Method:

- Use datasets from SDSS, DECaLS, Euclid, and JWST to gather redshift and rotational speed data of galaxies.
- Correlate redshift not with distance alone, but with galactic angular velocity, morphology, and ring structure.
- Identify anomalous redshifts between galaxies at similar distances but with different rotational characteristics.

Prediction (CMGED):

- Galaxies with slower or faster ω than the Milky Way will exhibit redshift or blueshift, regardless of distance.
- Redshift should not increase strictly linearly with distance, breaking from Hubble's law in favor of rotational divergence.

D.2 Ripple Signature Detection Pre- and Post-Merger

Objective: To identify gravitational ripple precursors or aftershocks from field-level tension realignments in galaxies.

Method:

- Re-examine LIGO/Virgo data and upcoming LISA missions for subtle, low-frequency gravitational ripples leading up to large mergers or collapse events.
- Cross-reference with electromagnetic precursors or infrared softening from high-tension galactic cores.

Prediction (CMGED):

- Ripple signals may precede or follow a gravitational wave event by seconds or minutes, depending on the scale of field rebalancing.
- These ripples will have radial propagation symmetry, not oscillatory spacetime curvature patterns.

D.3 Multi-Timeline Supernova Decay Rate Comparisons

Objective: To demonstrate that supernova light curves are

interpreted differently depending on the galactic time base of the observer and emitter.

Method:

- Use time-stamped light curves of Type Ia supernovae from galaxies with well-characterized rotation curves.
- Analyze decay rates, duration, and shape of the curves across redshift bands.
- Adjust interpretation using CMGED's angular time base framework.

Prediction (CMGED):

- Supernovae from galaxies with different ω will show non-uniform decay patterns that deviate from relativistic time dilation expectations.
- Anomalies in observed duration will correlate more strongly with rotational frequency difference than with redshift or distance.

D.4 Spiral Arm Regulation via Ripple Flow Mapping

Objective: To test if spiral arms act as energy regulators aligned with ripple feedback mechanics.

Method:

- Use radio telescopes and spectral line observations (e.g., HI 21 cm line) to map mass and energy flow along spiral arms of multiple galaxies.
- Observe density waves and star formation patterns in

coordination with field ripple propagation.

Prediction (CMGED):

- Spiral arms will show phase-aligned flow of mass and ripple wave coherence, acting as tension-dissipation pathways.
- Star formation regions will be concentrated at zones of ripple feedback realignment, not just density shocks.

D.5 Detection of Galactic "Breathing" through Subtle Periodic Motion

Objective: To observe periodic energy fluxes or mass migration within the Ro–Ri structure of galaxies, indicating field-regulated breathing cycles.

Method:

- Track radial mass distribution shifts and gas movement in galaxies over time using ALMA, VLT, and future SKA arrays.
- Monitor central mass inflow and peripheral adjustments to identify breathing rhythms.

Prediction (CMGED):

- There will be periodic, low-amplitude, symmetrical inward-outward flows of mass-energy within the disc.
- These cycles will correlate with ripple emissions and minor spectral changes in core emissions.

D.6 Simulation of Nested Rotational Structures

Objective: To simulate multi-layered rotational systems and assess whether rotational detuning alone can reproduce redshift-like effects and stability.

Method:

- Use numerical field simulation platforms (e.g., GADGET-4, AREPO) modified with CMGED angular feedback and SVE logic.
- Run simulations of nested Ro–Ri systems, monitor how angular speed divergence affects signal propagation.

Prediction (CMGED):

- The simulations will replicate redshifted signals between rotating rings without requiring spatial separation.
- Systems will self-regulate under ripple feedback, preventing collapse and mimicking observed galactic behaviors.

D.7 Time Base Divergence via Clock Drift Observation

Objective: To test if clocks aboard probes moving at different galactic positions or field densities begin to diverge not due to relativistic speed, but due to field-regulated angular frequency.

Method:

- Send synchronized atomic clocks aboard probes to regions of differing galactic radial distances or neighboring star systems.
- Track clock drift relative to Earth and compare with local

galactic rotation parameters.

Prediction (CMGED):

- Clock drift will correlate with differences in galactic angular speed ω rather than only velocity or gravitational potential.
- This will provide strong support for angular time base theory over relativistic time dilation.

Summary Table

Experiment	Tested Concept	Expected Outcome (CMGED)
Rotational Redshift Mapping	Redshift = ω mismatch	Anomalous redshift patterns independent of distance
Ripple Signature Detection	Gravitational waves = SVE correction	Pre/post-ripples detectable with radial symmetry
Supernova Light Curves	Time base mismatch	Variable decay rates across redshifts & ω
Spiral Arm Mapping	Arms as field regulators	Ripple-correlated star formation & flow
Galactic Breathing	Mass-energy cycles	Periodic inward-outward shifts
Rotational Simulations	Layered structure dynamics	Emergent redshift and balance without expansion
Clock Drift Tests	Angular time base	Drift linked to ω, not just speed or gravity

These experiments and observations are designed not to disprove relativity or classic models, but to validate a higher-order layer of understanding — one where structured fields, rotational consequence, and vacuum intelligence replace inertia, void, and gravitational mystery.

CMGED is ready to be tested. The sky already holds the evidence — we just need to ask the right question.

Appendix F

Preliminary Validation: The Andromeda–Triangulum Test Case

One of the fundamental claims of Chavan's Model of Galactic Energy Dynamics (CMGED) is that spectral shifts observed in galaxies—specifically redshifts and blueshifts—are not reliable indicators of linear motion through space (as in cosmic expansion), but rather signatures of rotational frame mismatch between galaxies embedded in Structured Vacuum Energy (SVE).

To test this claim, we evaluate two well-known nearby galaxies: Andromeda (M31) and Triangulum (M33). Both are part of the Local Group, gravitationally bound and not significantly influenced by any cosmic expansion, making them ideal candidates for testing CMGED's rotational redshift hypothesis.

F.1 Observational Data Summary

Galaxy	Distance from Earth	Rotational Velocity	Spectral Shift	Mainstream Interpretation	CMGED Interpretation
M31 (Andromeda)	~765 kpc (~2.5 million ly)	~250 km/s	Blueshift	Moving toward the Milky Way	Rotating faster than the Milky Way
M33 (Triangulum)	~849 kpc (~2.77 million ly)	~100 km/s	Redshift	Moving away locally	Rotating slower than the Milky Way

F.2 Interpretation According to CMGED

In CMGED, light emitted by a galaxy carries the imprint of that galaxy's rotational time base — defined by its angular velocity within its galactic Ro–Ri structure. When this light reaches another galaxy, its frequency is interpreted according to the observer's own rotational time base.

Thus:

- If the emitter rotates faster than the observer, the signal appears blueshifted.
- If the emitter rotates slower than the observer, the signal appears redshifted.

This model perfectly accounts for the observed behavior of M31 and M33:

- M31, with higher angular velocity, appears blueshifted.
- M33, with lower angular velocity, appears redshifted.

Despite both galaxies being part of the same gravitationally bound system (not expanding apart), their spectral shifts diverge — not due to recession, but due to rotational detuning.

F.3 Scientific Implications

This simple test case supports several critical assertions of CMGED:

- Redshift is not an absolute indicator of distance or recession.
- Rotational speed is a determining factor in spectral interpretation.
- Field-based, angular time frames offer a more coherent explanation than inertial motion or expanding metrics.

This validation encourages a systematic re-analysis of galactic redshifts not as Doppler effects, but as resonance mismatches in rotational systems. It opens the door for predictive modeling based on angular velocity, rather than radial velocity.

F.4 Proposed Title for Expanded Study

"Rotational Redshift: A Field-Based Redefinition of Cosmic Motion"
Validating the Chavan Model through the Andromeda–Triangulum Test Case

This appendix marks the beginning of CMGED's observational validation, demonstrating that the universe may not be expanding—but rather rotating in layers, and whispering its structure through the rhythm of light.

CMGED in a Nutshell

Chavan's Model of Galactic Energy Dynamics (CMGED)

Redefining Galactic Rotation, Redshift, and Collapse through Universal Energy Dynamics (UED)

What Is CMGED?

A field-based, consequence-driven cosmological model that replaces the speculative elements of modern astrophysics (dark matter, spacetime curvature, Big Bang expansion) with a mechanically grounded framework of structured vacuum fields, nested rotation, and ripple-regulated equilibrium.

Core Concept: The Ro–Ri Ring

Every galaxy exists as a rotating annular structure bounded by two key radii:

- Ri = Inner rotational limit (near core, high field tension)
- Ro = Outer rotational limit (coherence fades, halo begins)

All galactic motion, light, and memory are confined to this Ro–Ri rotational shell, regulated by structured vacuum energy (SVE).

What CMGED Explains Better:

Problem	CMGED Solution
Flat rotation curves	Constant angular velocity from field-based moment of inertia regulation
Redshift	Rotational time-base mismatch, not spatial recession
Gravitational waves	Ripple feedback from field imbalance, not space-time warps
Galactic mergers	Field-level resets when angular harmony fails
Collapse & black holes	Frictional sinks — not destructive singularities but structured absorption
Cosmic expansion	Illusion caused by rotational detuning across nested SVE fields

Key Mechanisms:

- **Structured Vacuum Energy (SVE):** The real medium — not space, not ether, but field intelligence.
- **Ripple Feedback:** Radial pulses that correct system imbalance (a reinterpretation of gravitational waves).
- **Consequence-Driven Motion:** Motion arises from field tension — not from inertial mechanics.
- **Rotational Time Bases:** Each system defines time by its spin — redshift arises from detuned rhythms.

Observational Anchors:

- Andromeda–Triangulum Test Case: Different redshift signatures align with rotational speed differences, not distance.
- Gaia & SDSS Data: Support nested co-motion, field-bound spin alignment, and rotation-regulated evolution.

Philosophical Shift:

- The universe is not expanding chaotically — it is rotating into structure.
- Stillness is not the absence of motion — it's the presence of perfect motion.
- Collapse is not destruction — it's field reabsorption into equilibrium.

If You Remember Just 3 Things:

1. Galaxies are not moving away — they are spinning out of sync.
2. Black holes don't destroy — they recycle.
3. The universe breathes — through rings, ripples, and balance.

4.

References

Primary Observational Missions and Data Archives

1. **Gaia Data Release 3 (2022)** *European Space Agency (ESA)*
    - Precision astrometric catalog of over 1.8 billion stars
    - Source for stellar motions, galactic structure, and co-motion analysis
    - URL: https://gea.esac.esa.int/archive/
2. **Sloan Digital Sky Survey (SDSS) – Data Release 17**
    - Comprehensive spectroscopic redshift survey of millions of galaxies
    - Provides spin data, galactic morphologies, and large-scale structure
    - URL: https://www.sdss.org/dr17/
3. **LIGO Scientific Collaboration and Virgo Collaboration (2016–Present)**
    - Detection of gravitational waves from binary mergers
    - Data accessed via Gravitational Wave Open Science Center
    - URL: https://www.gw-openscience.org/
4. **NASA/IPAC Extragalactic Database (NED)**
    - Centralized database of galaxy redshifts, distances, and rotational profiles
    - URL: https://ned.ipac.caltech.edu/
5. **Galaxy Rotation Curves Archive (Courteau et al.)**
    - Compilation of observed rotation curves for spiral galaxies
    - Essential for mapping angular velocities and mass

profiles
  - Referenced in: Courteau, S. (1997). *AJ, 114, 2402*

Theoretical and Supportive Papers

1. Milgrom, M. (1983). *A modification of the Newtonian dynamics – Implications for galaxy systems.* ApJ, 270, 371.
   - Provided initial motivation to question dark matter by exploring alternate dynamics.
2. McGaugh, S. S., Lelli, F., & Schombert, J. M. (2016). *Radial Acceleration Relation in Rotationally Supported Galaxies.* Physical Review Letters, 117(20), 201101.
   - Demonstrates empirical correlations inconsistent with dark matter halos.
3. Sellwood, J. A., & Carlberg, R. G. (1984). *Spiral structure in disk galaxies.* ApJ, 282, 61.
   - Supports ripple-like nature of spiral arms, consistent with CMGED field regulation.

Cosmological Overviews and Datasets

1. Freedman, W. L., & Madore, B. F. (2010). *The Hubble Constant.* ARA&A, 48, 673.
   - Useful baseline for contrasting CMGED's reinterpretation of redshift.
2. Peebles, P. J. E., & Ratra, B. (2003). *The Cosmological Constant and Dark Energy.* Reviews of Modern Physics, 75(2), 559.

- Theoretical foundation of Lambda-CDM model, contrasted by CMGED.

Specific Galaxy Case Studies

1. **Andromeda Galaxy (M31)** – Wikipedia contributors. (2024). *Wikipedia, The Free Encyclopedia.*

- Rotational speed and blueshift confirmation.
- https://en.wikipedia.org/wiki/Andromeda_Galaxy

1. **Triangulum Galaxy (M33)** – Wikipedia contributors. (2024). *Wikipedia, The Free Encyclopedia.*

- Used for redshift and rotational comparison.
- https://en.wikipedia.org/wiki/Triangulum_Galaxy

Software and Simulation Resources

1. Springel, V. (2005). *The cosmological simulation code GADGET-2.* MNRAS, 364(4), 1105.

- Referenced for CMGED simulation proposals.
- GADGET-4 is the latest platform for large-scale field simulations.

1. Open Source Field Modeling Tools

- For ripple propagation, angular momentum modeling, and nested structure simulations:
    - **yt-project** (https://yt-project.org)
    - **PyGalaxies** (custom package)

Authorial Works

1. Chavan, S. (2025). *Universal Energy Dynamics: Foundations of Structured Vacuum and Consequential Motion*

2. Chavan, S. (2025). *The Law of Consequence and Rotational Equilibrium in the Cosmos*
3. Chavan, S. (2025). *CMGED: A Field-Based Reinterpretation of Galactic Dynamics and Redshift* (in preparation)

Acknowledgment

Observational Truth – Sky as the Primary Source

Before telescopes, before equations, before theories—there was the sky.

In the development of Chavan's Model of Galactic Energy Dynamics (CMGED), while scientific literature and astronomical data were valuable, the primary source of insight was not a paper or a simulation, but observation—the persistent, patient act of watching the sky with an open and questioning mind.

This work acknowledges a fundamental principle often lost in modern science:

The cosmos is not revealed in laboratories alone—it reveals itself in silence, rhythm, and light.

The Sky Is the First Teacher

The fixedness of constellations, the quiet spiral of galaxies, the spectral whispers carried in starlight—these are not merely aesthetic curiosities. They are data of the deepest kind, encoded in patterns visible to anyone who dares to pause long enough to see.

It is through such reflection that the foundational concepts of CMGED were born:

- The shared motion of stars, perceived not as inertia but as co-resonance within a structured medium.
- The realization that redshift may not mean distance, but

difference in rhythm.
- The truth that gravitational waves are not forceful ruptures, but gentle corrections from a field that prefers balance.

These ideas were not first written—they were first felt, then tested, then structured.

A Tribute to Direct Observation

This model was not conceived in isolation from the universe—it was conceived in direct dialogue with it. It was shaped by:

- Watching the stars retain their positions over 40 years,
- Noticing that movement requires contrast, and contrast demands a still reference,
- Seeing that stillness itself is a form of perfect motion, synchronized across an invisible frame.

Thus, the night sky was not background—it was collaborator.

Gratitude to the Natural Source of Truth

This acknowledgment is for the sky itself—the grand mirror of consequence and the most honest record-keeper of cosmic behavior. It has never lied, never exaggerated, never conformed to models. It simply moves as it must.

To the stars, which rotate but do not flee.

To the darkness, which holds structure, not void.

To the ripples, which correct without force.

And to the silence, which reveals what the noise could not...

This work bows in gratitude.

Let this be a reminder: We do not invent the universe. We remember it. And the sky is the script where memory is written.

Further Reading

UED, CMPED, and Related Works

The ideas presented in *Chavan's Model of Galactic Energy Dynamics (CMGED)* are not isolated hypotheses but part of a larger, interlinked intellectual framework that spans cosmology, wave theory, black hole physics, consciousness, and structured energy systems. The following selected works provide essential context, depth, and extended discussion for readers seeking to explore the foundations and implications of CMGED.

Foundational Reports and Scientific Research (Available on Zenodo)

1. *Chavan's Law: A Universal Energy Equilibrium Model for Black Holes*

DOI: 10.5281/ZENODO.14782900, Establishes the equilibrium logic that underpins energy behavior around and within black holes.

1. *Formation of Black Holes: A Universal Energy Dynamics (UED) Perspective*

DOI: 10.5281/ZENODO.14788539, Reframes black holes as frictional sinks within a structured vacuum field.

1. *The Myth of Wave-Particle Duality: A Structured Vacuum Energy Perspective*

DOI: 10.5281/ZENODO.14793162, Challenges quantum orthodoxy by presenting light and matter as pure structured energy disturbances.

1. *Chavan's Structured Energy Equilibrium Principle (SEEP)*

DOI: 10.5281/ZENODO.14916491, Introduces the equilibrium dynamics of rotating systems, central to CMGED.

1. *Chavan-Einstein Unified Theory (CEUT): A Unified Framework for Gravity, Matter, Energy, and Consciousness*

DOI: 10.5281/ZENODO.14992289, Bridges structured vacuum fields with general relativity and deeper metaphysical insights.

1. *Chavan's The Law of Consequences (LOC)*

DOI: 10.5281/ZENODO.15048775, Foundation of motion and feedback in the CMGED and broader UED frameworks.

1. *Chavan's UED: Universal Framework of Realization and Perception*

DOI: 10.5281/ZENODO.15105060, A philosophical and scientific synthesis defining UED's broader implications for reality.

1. *Chavan's Model of Planetary Energy Dynamics (CMPED)*

DOI: 10.5281/ZENODO.15164927, Describes field-embedded motion of planets, forming the precursor logic to CMGED.

1. *The New Wave Theory: Chavan's Structured Resonance Framework (CSRF)*

DOI: 10.5281/ZENODO.15164881, Develops the concept of resonance fields in place of conventional particle models.

Books Related to the UED Framework

- *Unified Energy Dynamics:* The umbrella theory connecting vacuum structure, consequence, and equilibrium.
- *The Blueprint of the Universe:* Lays the groundwork for understanding the structured energy order beneath all phenomena.
- *The Cosmic Breath:* Explores the idea of galaxies "breathing" in rhythmic harmony with the vacuum field.
- *Physics Fixing:* A critical commentary on where mainstream physics needs correction—and how UED offers it.
- *The Savior Illusion & The Absorption:* There Is No Death: Blend metaphysical interpretation with field physics, showing continuity of energy beyond collapse.
- *The Fundamental Theory of Life & Existence (Series):* Titles such as *Energy in Action*, *The Active Mind*, *Death: The Disconnection* reinforce the continuity of UED principles across life, mind, and cosmic death.

Philosophical and Reflective Extensions

- *You Are Just a Version of Your Original*

- *Take It Easy: The Simplest Solution for Any Problem*
- *The Future Knows No Lines*
- *Human Side of Einstein*
- *Neural Wars: War Against Absolute Logic*

These works extend UED thinking into the human realm—consciousness, technology, identity, and perception—showing how field dynamics influence even the most subjective aspects of life.

Online Portal for Publications & Collaborations

Visit www.sandeepjchavan.com[1] for:

- Complete list of publications
- Updates on UED and CMGED research
- Collaboration opportunities
- Contact and discussion forums

This further reading list is not only a continuation of CMGED—it is its ecosystem. Every title contributes a piece to the evolving puzzle of a universe that spins not for survival, but for equilibrium, resonance, and silent intelligence.

1. https://www.sandeepjchavan.com

Disclaimer

This document presents the *Chavan's Model of Galactic Energy Dynamics (CMGED)* as a theoretical and conceptual framework developed by Sandeep Chavan under the broader paradigm of Universal Energy Dynamics (UED). The model is intended to offer a field-based alternative to mainstream cosmological interpretations, such as the Big Bang theory, spacetime curvature, and dark matter/dark energy constructs.

All observations, reinterpretations, and conceptual frameworks herein are derived from:

- Direct empirical references,
- Public scientific data sets (e.g., Gaia, SDSS, LIGO),
- Independent research and publications authored by Sandeep Chavan.

This work is not affiliated with any governmental, academic, or institutional scientific body and does not claim conformity with prevailing standard models in cosmology or physics. It should be read as an independent scientific contribution and alternative hypothesis, intended to provoke inquiry, stimulate debate, and offer new directions for theoretical and observational exploration.

The content should not be interpreted as absolute truth, conclusive evidence, or replacement for rigorous peer-reviewed consensus, but rather as a philosophically aligned, physically reasoned framework grounded in observable patterns and logical consistency.

Readers, researchers, and institutions are encouraged to critically evaluate the ideas, replicate suggested observations, and apply scientific scrutiny to every claim made. The author welcomes constructive dialogue, challenges, refinements, and collaborative inquiry.

The universe is not obligated to match human theories. But human theories must continuously evolve to match the universe.

Don't miss out!

Visit the website below and you can sign up to receive emails whenever Sandeep Chavan publishes a new book. There's no charge and no obligation.

https://books2read.com/r/B-A-EPGPC-TSZFG

BOOKS 2 READ

Connecting independent readers to independent writers.

Did you love *CMGED Research Report*? Then you should read *The Cosmic Breath*[1] by SANDEEP CHAVAN!

[2]

The universe is not a static entity; it breathes, evolves, and transforms in cycles of creation and dissolution. In ***The Cosmic Breath: Black Holes, Pralaya, and the Energy Cycles of the Universe*, Sandeep Chavan** explores this eternal cosmic rhythm, drawing compelling connections between modern astrophysics and ancient Vedic cosmology to offer a profound new perspective on the structure and fate of the universe.

At the heart of this exploration lies the mystery of black holes—once considered cosmic endpoints, now emerging as potential regulators of energy and time. Chavan challenges the singularity paradigm, proposing instead that black holes function as structured

1. https://books2read.com/u/380z5r
2. https://books2read.com/u/380z5r

energy equilibria, mirroring the Vedic concept of Pralaya, the dissolution phase that precedes cosmic rebirth.

Through ten interconnected chapters, the book delves into:

The Cyclic Nature of the Universe – How Pralaya aligns with modern theories like the Big Crunch, Big Bounce, and cosmic inflation.**Black Holes as Cosmic Portals** – Exploring whether black holes are not just destructive forces but key to recycling energy across space-time.**Time and Reality** – Investigating time distortion near black holes and its resemblance to Kaal Bhairav, the Hindu deity representing the end of time.**Dark Energy and Cosmic Expansion** – Examining how dark energy may not be an unknown repulsive force but a structured energy flow, echoing the ancient idea of Prana (universal energy).**The Future of Space-Time Science** – Challenging the mechanistic view of the cosmos and advocating for a new paradigm that integrates metaphysics and physics.

By bridging ancient insights with modern discoveries, *The Cosmic Breath* invites readers to rethink their understanding of space-time, energy, and the very fabric of existence. It questions whether the universe follows an infinite cycle of expansion and contraction—a cosmic breath ensuring perpetual renewal.

Beyond scientific exploration, Sandeep Chavan presents a philosophical journey, encouraging curiosity, open-mindedness, and interdisciplinary thinking. This book urges readers to look beyond conventional physics and consider whether cosmic evolution, energy balance, and consciousness are interconnected forces shaping the universe.

With an engaging, fluid narrative, *The Cosmic Breath* is for scientists, philosophers, and seekers alike—anyone drawn to the mysteries of black holes, the origins of time, and the grand cycles of the cosmos.

Read more at https://www.sandeepjchavan.com/.

Also by Sandeep Chavan

Echoes of War: The Russia-Ukraine Conflict and Its Global Ripples

Shattered Lives
The War Economy
Borders Redrawn
Arms and Allies
Collateral Damage
Justice in the Ashes

Research Reports – UED

CMGED Research Report

Somewhere in the Darkness of the Cosmos

The Blueprint of the Universe
The Cosmic Breath

The Eternal Sage Chronicles

The Sage and the Sovereign

Ultra Modern Life Philosophy

The Duality of Laughter and Tears

Take It Easy

Compare to Accept

Standalone

It's Not AI, It's AHI - Amplified Human Intelligence

The Decision Paradox

The Four Sapiens

Malicious Script of Indian Polity

The IIT Legacy & Global Impact

The IIT Legacy & Global Impact

You are just a Version of Your Original

Win the Game You Didn't Choose

Always Keep Your Bags Packed

Reveal to Shield - Learn the Game of Facades

Engage Beyond Elections

Culture, Identity & Change: The Evolution of Indian Society

The Triangular Dynamics

Never Trust Without Doubt

If You Can't Change the Country, Leave the Country

Live Fearless with Fear

India: The Investment Magnet

Realignment

Powering a Billion Dreams

Universal Skill

Extension: From Brush to Algorithm

Human Side of Einstein

The Future Knows No Lines

Let's Bring the Generations to the Table
Unified Energy Dynamics
A Beat That Never Dies
Strategic Horizons
Right From Scratch
The Forgotten Innovators
Physics Fixing
The Savior Illusion
The Selfish Side
The Age of the Weakest Sapiens
Neural Wars
The Absorption: There Is No Death
Strange Therapist
Bubbles: Become a Drop
Beyond the Blue
Half-Scholar
The Law of Truth
Failure

Watch for more at https://www.sandeepjchavan.com/.

www.ingramcontent.com/pod-product-compliance
Lightning Source LLC
LaVergne TN
LVHW090419210926
848480LV00002B/620